Beef Cattle Breeds For Beef and For Milk
Farmers' Bulletin No. 1779

by US Dept. of Agriculture

with an introduction by Jackson Chambers

Self Reliance Books

Get more historic titles on animal and stock breeding, gardening and old fashioned skills by visiting us at:

http://selfreliancebooks.blogspot.com/

Introduction

I am pleased to present another title in the "Cattle" series.

The work is in the Public Domain and is re-printed here in accordance with Federal Laws.

As with all reprinted books of this age that are intended to perfectly reproduce the original edition, considerable pains and effort had to be undertaken to correct fading and sometimes outright damage to existing proofs of this title. At times, this task is quite monumental, requiring an almost total "rebuilding" of some pages from digital proofs of multiple copies. Despite this, imperfections still sometimes exist in the final proof and may detract from the visual appearance of the text.

I hope you enjoy reading this book as much as I enjoyed making it available to readers again.

Jackson Chambers

BRIEF descriptions of the breeds of cattle that are kept for beef or for both beef and milk in the United States are presented in this bulletin.

A breed is a race or variety of animals related by descent and similar in most characters.

To preserve the pedigrees and to promote the various breeds, registry associations have been established for most of the breeds described in this bulletin. The authors gratefully acknowledge the cooperation of the various breed association secretaries in providing written material and illustrations.

The names and addresses of the secretaries of these associations may be obtained upon request from the Animal and Poultry Husbandry Research Branch, United States Department of Agriculture, Washington 25, D. C.

Revised January 1954

Washington, D. C.

Slightly revised March 1958

CONTENTS

Beef-Cattle Breeds for Beef and for Beef and Milk

By R. T. CLARK and A. L. BAKER,[1] *Animal Husbandry Research Division, Agricultural Research Service*[2]

DEVELOPMENT OF BEEF CATTLE BREEDS

MOST of the initial developmental work on beef cattle breeds started in the middle of the 18th century.

The important beef breeds in the United States originated in Great Britain. The most prominent early British breeders were Robert Bakewell, the Colling brothers, Richard Tompkins, Amos Cruickshank, and Hugh Watson. These men are credited with the first constructive efforts to establish breeds and improve beef cattle in Great Britain. They placed considerable emphasis on records of production and on carcass values, and selected and bred for characters of a utilitarian nature. Subsequently, and for too long a time, their methods were not considered necessary.

With the accumulation of research information, we are witnessing a decided change in beef cattle breeding practices—a return to and an appreciation of the precise methods used by the early improvers of beef cattle.

During the latter part of the 18th century, the United States began importing representatives of the principal British beef cattle breeds. In general, these beef cattle breeds can be distinguished from other types of cattle by their true beef form and marked fleshing qualities.

In addition to having desirable beef form, the true dual-purpose breeds also produce higher levels of milk and butterfat than is usual in typical beef cattle. For these reasons dual-purpose cattle have been favored in certain farming areas in the United States.

Throughout the history of beef cattle breeds, considerable emphasis has been placed on selection of types that are rectangular in form, wide and deep, with depth of fleshing especially in the regions of the back, loin, rear quarters, and ribs.

Allowing for inherent and environmental differences in most of the beef cattle breeds, mature bulls should weigh 1,600 to 2,100 pounds, and mature cows from 1,100 to 1,400 pounds.

BREEDS DEVELOPED IN THE BRITISH ISLES
ABERDEEN-ANGUS

The Aberdeen-Angus breed originated in adjacent counties—Aberdeenshire and Forfarshire (now Angus)—in Scotland, from two local bovine strains known as humlies and doddies. The breed was developed under rather rigorous environmental conditions on land which is largely rolling to rough and not particularly fertile except in the valleys.

---

[1] Deceased.
[2] Earlier editions of this bulletin were written by W. H. Black, now deceased.

1

FIGURE 1.—Aberdeen-Angus bull.

FIGURE 2.—Aberdeen-Angus cow.

Early improvers of the breed in the United States credit George Grant, a native of Banffshire, Scotland, and a retired London silk merchant then living in Victoria, Kans., with the first importation of four Angus bulls in 1873. These bulls were crossed with native Texas cattle of the Longhorn variety. They adapted themselves to the range and worked so much improvement on the native cattle that further importations were made. In 1883, 14 steers sired by Grant's bulls were sold on the market at the Kansas City stockyards and created a favorable impression. Further importations of Aberdeen-Angus into the Corn Belt States increased the popularity of the breed in this area. Today the number of established Aberdeen-Angus herds is increasing rapidly and the American Aberdeen-Angus Breeders' Association reports that more purebred calves are being registered from the States west of the Mississippi than to the east. This is due to the wide acceptance of the breed in the range areas. The breed is found in every State and in Canada.

Aberdeen-Angus cattle are distinguished from other breeds by their black color, comparatively smooth coats of hair, and polled heads (figs. 1 and 2). These cattle are good rustlers and are able to adapt themselves rapidly to the varied climatic and grazing conditions throughout the country. They show marked resistance to eye troubles. This has been credited to the black pigment in the skin. They cross well with cattle of other breeds. The bulls possess a high degree of ability to transmit the polled characteristic and black color to their offspring. At least 95 percent of the calves sired by Aberdeen-Angus bulls and out of horned, domestic-type cows are hornless and black in color. After three to five generations of continued top crossing with Aberdeen-Angus bulls, practically the entire herd will be black and hornless.

The Aberdeen-Angus are bred and raised primarily for beef purposes. However, many strains within this breed are fairly good milkers, and it is only rarely that a cow of this breed fails to give sufficient milk for the proper development of her calf.

Aberdeen-Angus cattle hold an enviable record in the feed lot and as fat slaughter cattle. This is due primarily to their inherent ability to mature at an early age and to produce carcasses of high quality meat coupled with a high dressing yield. Marbling of the fat with the lean meat is also a characteristic of this breed.

The American Aberdeen-Angus Breeders' Association, formed in 1883, published volume I of the herdbook in 1886. The first secretary-treasurer of the registry group was Charles Gudgell, Pleasant Hill, Mo., who played an important role in the importation of large numbers of foundation beef cattle to this country.

In 1888, membership in the association, which had originally been fixed at 200, was made unlimited. In 1952, the active membership numbered 24,157. Total registrations at that time numbered approximately 1,600,000, and 82 volumes of herdbooks had been published.

The principal Aberdeen-Angus families are: Miss Burgess, Erica, Pride of Aberdeen, Queen Mother, Blackbird, and Blackcap. These families frequently had been broken up into strains, tribes, or branches named after particular females descending from the original cows.

DEVON

The Devon breed was developed in and derives its name from the county of Devon, which lies between the Bristol and English Channels in southwest England. This part of England is rolling to rough in topography, with soils that are considered only fair in fertility. The climate, though fairly equable, is damp and chilly. For these reasons, the Devon is not as large in body size as some of the other British breeds.

FIGURE 3.—Devon Bull.

FIGURE 4.—Devon cow.

The American Devon Cattle Club records show that Devons first came to North America 1623, when Winslow, agent for the Plymouth Colony, received a shipment of red cattle on the *Charity*.

Later importations were made into Massachusetts by Winthrop and Davenport as early as 1800, and into New York in 1805 by General Eaton. Soon afterward, other importations were made into Maryland and the New England States.

Today the greatest number of Devons are in the South—Texas, Mississippi, Louisiana, Florida, and Alabama being the States with the most herds. Probably three-fourths of the cattle of this breed are in the Gulf States. Cattle showing Devon blood appear to thrive in this humid region.

The Devon is red in color and is generally horned, although polled Devons also exist and can be registered. The horns are white or waxy, with dark tips. In bulls the horns are straight (fig. 3), and in cows they curve upward, forward, outward, and eventually backward (fig. 4). Devon cows are considered to be very good milk producers, and it is claimed that they are capable of producing young, milk-fat baby beef off grass economically.

Volume I of the American Devon Record was published in 1881 by James Buckingham, of Zanesville, Ohio. In 1905, when the American Devon Cattle Breeding Association was formed, the American Devon Record became the official register. The association is now organized as the American Devon Cattle Club, Inc.

GALLOWAY

The Galloway breed was developed in southwest Scotland in a district that included the counties of Kirkcudbright and Wigtown, originally known as the Province of Galloway. This breed is considered to be one of the oldest of British breeds. Except for Scotch Highland cattle, probably no British breed was ever selected under a more rigorous climate, for the district of Galloway is known for its dampness and cold. Much of this part of Scotland is rough, but in spite of that it is generally recognized as a good grazing section.

Some Galloways were introduced into the United States during the first half of the 19th century. During 1870 an importation was made into Michigan and later the breed was distributed throughout the Corn Belt States.

Galloways are black in color and are polled. In general appearance they resemble the Aberdeen-Angus. However, they may be distinguished from this breed by their characteristic curly hair coat (figs.

FIGURE 5.—Galloway bull.

FIGURE 6.—Galloway heifer.

5 and 6). Also, they exhibit a more rounded poll. This breed has been developed for the express object of breeding a high quality beef animal that is capable of adaption to rigorous grazing conditions. The breed has made a creditable showing in carcass contests.

The American Galloway Breeders' Association was organized in Chicago on November 23, 1882.

HEREFORD

The Hereford breed originated in England in the county of Hereford. The locality known as Herefordshire, lying between the Severn River and the eastern boundary of Wales, has much fertile valley and plains land which produces pastures and crops abundantly. The excellent pastoral and climatic conditions have favored the development of the breed.

The earliest importations of Hereford cattle into the United States, of which there is authentic record, were those made by Henry Clay and Lewis Sanders of Kentucky, in 1817, and those given the Massachusetts Society for the Promotion of Agriculture by Admiral Coffin of the Royal British Navy about 1825. These were soon followed by numerous importations.

The American Hereford Association reports that the first breeding herd of registered Herefords established in the United States was that of William H. Sotham and Erastus Corning of Albany, N. Y., in 1840.

The breed increased rapidly in favor and now has wide geographic distribution in the United States.

The Hereford is strictly a beef breed and has met with unusual favor in the range areas of the West.

The Hereford is readily distinguished from all other breeds by its color markings—red body and white face. The white color is found also on the underline, flank, crest, switch, breast, and below the knee and hock (figs. 7 and 8).

FIGURE 7.—Hereford bull.

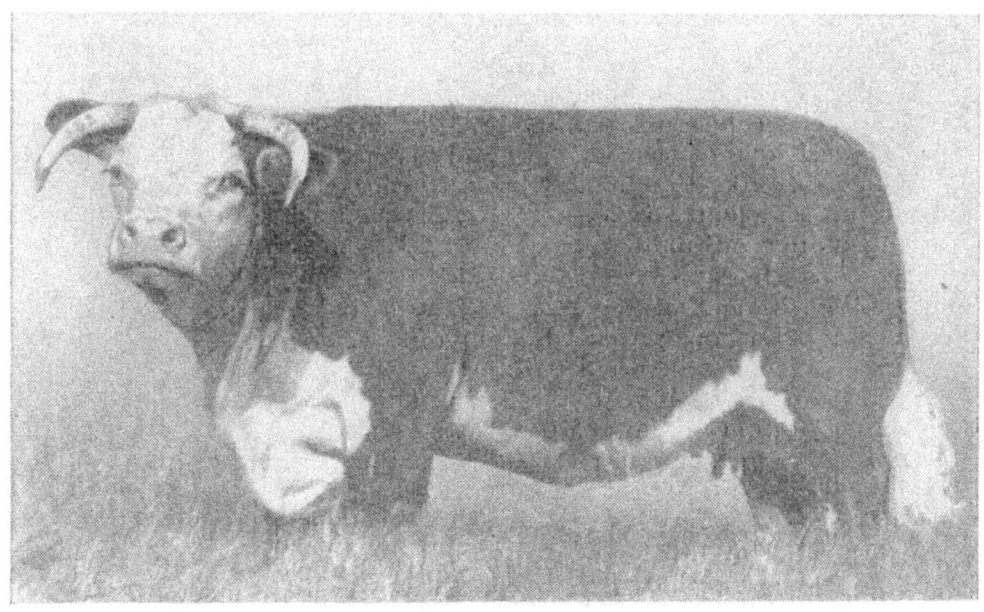

FIGURE 8.—Hereford cow.

As Herefords have been raised exclusively for beef purposes, milking qualities have generally not been stressed.

The American Hereford Cattle Breeders' Association was organized in 1881 and published in that year volume I of the American Hereford Record. The official name of the Association was changed on October 22, 1934, to the American Hereford Association.

The following sires are among those which have contributed much to the development of the breed in the United States: Anxiety 4th 9904, Don Carlos 33734, Beau Brummel 51817, Beau Donald 58996, Perfection Fairfax 179767, Domino 264259, Repeater 289598, Gay Lad 16th 316946, Beau Blanchard 362904, Gay Lad 9th 386873, Bocaldo 6th 464826, Prince Domino 499611, Woodford 500000, Braemore 666666, Panama 100th 786758, Dandy Domino 2nd 1090962, Hazford Tone 1093542, Mischief Mixer 27th 1179215, Beau Blanchard 155th 1202407, Hazford Rupert 25th 1209734, Prince Domino 2nd 1222880, Hartland Mischief 1314000, Prince Domino Mixer 1458747, Prince Domino C. 1565007, WHR Royal Domino 2nd 1849068, and Larry Domino 50th 2624412.

POLLED HEREFORD

The Polled Hereford was developed in the United States by Warren Gammon of Des Moines, Iowa. In 1900, he circularized Hereford breeders and obtained 13 head of polled mutants that were registered or eligible for registry in the American Hereford Record. Eleven head were registered in 1901, when the Record was established. From this nucleus, and through the use of polled sires on horned females, the population of Polled Herefords has increased rapidly. They are now in demand all over the world, particularly in areas where external parasites have to be reckoned with. The five States leading in production of Polled Herefords are Texas, Illinois, Kansas, Nebraska, and Missouri.

The American Polled Hereford Association maintains its own Record, with the stipulation that no animal is eligible for registration

FIGURE 9.—Polled Hereford bull.

unless it traces in every line to animals recorded in volume XIII or prior volumes of the English Hereford Herd Book.

RED POLL[3]

The Red Poll breed originated in the eastern middle coastal section of England by the crossing and eventual merging of the native inbred stocks of Norfolk and Suffolk cattle. The merger started in the early years of the 19th century, and was virtually complete by 1850. The breed was recognized by the Royal Agricultural Society before 1850. The first English herdbook was published as a private enterprise by Henry F. Euren in 1874. The Red Poll Cattle Society of Great Britain and Ireland was organized in 1888 and took over the records collected by Euren. The Red Poll Cattle Club of America was organized in 1883 and published the first American volume of the herdbook in 1888. All American registered Red Polls today trace to those registered in volume I of the English herdbook.

Although a polled Suffolk heifer was sent from Ireland to the Jamestown colonists as a gift in early colonial times and founded a strain of native cattle known as the "Jamestown" cattle, the first authentic importations of registered Red Polls were made by G. F. Taber, of New York, in 1873, 1875, and 1882. Others followed and approximately 300 head were imported by 1900. Small importations of 1 to 13 head were made in 1929, 1941, 1949, and 1950. Some of these had considerable influence on recent improvement of the breed, others remain to be proved.

Red Poll cattle are, as the name indicates, both red and polled. The preferred color is a deep, rich cherry red. Variations from light

[3] Official name of breed changed from Red Polled to Red Poll in 1946.

FIGURE 10.—Red Poll bull.

FIGURE 11.—Red Poll cow.

red to very dark red exist. Natural white is normal in the switch of the tail and is permitted below the underline in limited amounts.

Red Polls of good type are moderately short of leg, deep in the body, moderately thick, and have a smooth, even flesh covering (figs. 10 and 11). At maturity, in medium flesh, the bulls attain weights of 1,800 to 2,000 pounds, cows 1,200 to 1,500 pounds.

Red Polls since their origin have been dual purpose and have proved themselves to be good farm cattle, as the cows are capable of producing comparatively large quantities of milk and the calves of growing into desirable beef animals. The breed is considered early maturing. Red Polls are most numerous in the Midwest, but are increasing in the East, in the irrigated areas of the West, and in the South. In the South they are attaining popularity as beef cattle. There the bulls are crossed with native cows and with cows of pure beef breeds.

The 10 top breed-improving sires in 1953 were: Elgin Advancer 56316 AR; Kirton Statesman (Imp.) 63325 AR; Fairfax Charmer 57572; Gold Coin Advancer 61203 AR; Redvue Norman 58759 AR; Melbourne Advancer 60366 AR; M. L. F. Pat Again 58576 AR; Nora's Charmer 61230 AR; St. Patrick 56494; Paydown S. Leona's Pattern 6115 AR.[4]

SCOTCH HIGHLAND

The Scotch Highland breed of cattle, developed in the Hebrides, is reputed to have been recorded as a definite breed in the 12th century. In size or body weight, it is considered one of the smaller or, at most, intermediate breeds. Owing to continued selection under adverse

[4] Data for advanced registry were obtained by giving relative point values for each appearance of each sire in the three-generation pedigrees of the animals concerned.

FIGURE 12.—Scotch Highland bull.

conditions involving rigorous climate and at times scant feed supplies, animals of this breed have inherent hardiness coupled with mothering ability. Acceptable colors are black, brindle, red, light red, yellow, dun, and silver. The horns are wide and branching (figs. 12 and 13). The ears are quite often short. The hair coat has been appropriately described as a "coat and vest," the undercoat being soft and thick while the outer coat is coarse and long. They have straight lines and in their original home they are reputed to produce excellent carcasses of high quality.

The American Scotch Highland Breeders Association was formed in 1948.

FIGURE 13—Scotch Highland cow.

SHORTHORN

The Shorthorn breed originated in the counties of Durham, Northumberland, and York, in northeastern England, and particularly in the valley of the Tees River. This valley has excellent pastures conducive to the development of large beef breeds.

The first importation of Shorthorns into the United States was made in 1783 by Miller and Gough of Virginia and Maryland. Early importations of Shorthorns were sometimes referred to as Durham, but this term is obsolete today.

Lewis Sanders of Kentucky imported Shorthorns in 1817, and in 1853 Samuel Thorne of New York and Abram Renick of Ohio imported foundation animals. These two breeders are often referred to as the founders of the Shorthorn breed in the United States.

Shorthorns are either roan, red, red with white spots, or white in color. They are rectangular in shape, and are one of our largest breeds of beef cattle (figs. 14 and 15).

Prior to 1883, there were three separate herdbooks for Shorthorn cattle in the United States. In that year Shorthorn breeders, holding their first national convention, decided to consolidate all registrations in one herdbook and the first consolidated volume, No. 24, was issued in 1883.

Three types of Shorthorn cattle are now recognized in the United States—Beef Shorthorn, Milking Shorthorn, and Polled Shorthorn.

BEEF SHORTHORN

Beef Shorthorns are particularly adapted to farming areas where there is an abundance of feed, for, owing to their size, Shorthorns are capable of using large quantities of roughage. Shorthorns cross well with other breeds for the production of commercial cattle. The

FIGURE 14.—Beef Shorthorn bull.

FIGURE 15.—Beef Shorthorn cow.

breed is well distributed throughout the United States, but the greatest numbers are in the Corn Belt. Shorthorn cows are usually good milkers, and therefore produce rapid-growing calves.

According to the American Shorthorn Breeders' Association, the following bulls are recognized as outstanding in breed history: Whitehall Sultan 163573; Avondale 245144; Browndale Count 1156438; Edellyn Campion Mercury 2071109, and Edellyn Royal Leader 2057560.

MILKING SHORTHORN

In the United States the strains of Shorthorn cattle that have been selected for both milk and beef production are known as "Milking Shorthorns." (Figs 16 and 17.) In England and Australia they are called "Dairy Shorthorns."

Although the primary objectives of the breeders of this type of Shorthorn have been to develop cattle that will produce moderately large quantities of milk, and steer calves that would finish out as acceptable beef, it is commonly thought that greater emphasis in selection has been placed on the milking qualities than on the beef-making potentials. Regardless of this relative emphasis, Milking Shorthorn steers are capable of yielding carcasses that are acceptable to perhaps all but the most discriminating beef trade. The Milking Shorthorn has been used to a large extent under general farming conditions, particularly in the Corn Belt and Eastern States.

While the majority of Milking Shorthorns are horned, there are a number of herds of Polled Milking Shorthorns.

The American Milking Shorthorn Society was formed in 1920. Since January 1, 1951, no animals have been accepted for registration in the American Milking Shorthorn Herd Book unless both sire and dam were already registered in that herdbook or have certificates of

FIGURE 16.—Milking Shorthorn bull.

registration as Milking Shorthorns issued prior to September 1, 1949, by the American Shorthorn Breeders' Association.

Under appropriate rules of the Society, a grading-up register is maintained leading to final purebred registration. This is intended to assist farmers to build up grade herds through the use of registered Milking Shorthorn bulls.

Some of the sires most noteworthy in the early development of the Milking Shorthorn in the United States are: Royal Knight 448999; General Clay 255920: Flintstone Gift 807690: Balthazar 614650; Darlington Duke 1306326: Glenside Dairy King 443881.

FIGURE 17.—Milking Shorthorn cow.

POLLED SHORTHORN

The early history of this breed is quite clear in contrast to that of the older breeds. The Polled Shorthorn (figs. 18 and 19) was developed principally in the Corn Belt States of Ohio, Indiana, Illinois, and Minnesota, in the late eighties. W. S. Miller, Elmore, Ill., in 1873, established a herd of polled cattle. In 1888, he purchased from a

FIGURE 18.—Polled Shorthorn bull.

FIGURE 19.—Polled Shorthorn cow.

breeder in Minnesota the polled twins, Mollie Gwynne and Nellie Gwynne, their half brother, King of Kine, and two yearling heifers. All of these animals were polled and registered in the Shorthorn Herdbook. King of Kine, when bred to his half-sister, Nellie Gwynne, produced Ottawa Duke. This bull sired a large number of polled progeny that appear in the initial volumes of the Polled Durham Herd Book. The American Polled Durham Association was organized in Chicago on November 14, 1889.

The use of the term "Polled Durham" was discontinued in 1919 in favor of "Polled Shorthorn," and in 1923 the Polled Shorthorn Association was disbanded. On January 1, 1923, Polled Shorthorns became eligible to registry in the American Shorthorn Herd Book.

J. H. Miller of Peru, Ind., is credited with making a significant contribution to the development of the Polled Shorthorn.

SUSSEX

The Sussex breed originated in the counties of Sussex and Kent in southern England. It is considered to be one of the oldest breeds developed in Great Britain. Originally, Sussex cattle were used for draft purposes, which undoubtedly accounts for their relatively large size and good feet (fig. 20).

FIGURE 20.—Sussex cow.

The Sussex is noted for its high average daily gain. Typical representatives are horned, and dark red in color.

Outside of its country of origin, the Sussex has attained its greatest prominence in the Union of South Africa, in southwest Africa, and in Rhodesia. It has crossed well with the native Africander. Only a small number of Sussex have been introduced into the United States.

BREEDS INTRODUCED FROM OTHER COUNTRIES
BRAHMAN

The cattle of India (*Bos indicus*) are called "Brahman" in the United States and are described as "Zebu" in Europe and South America (fig. 21). More than one-half of all the cattle in the world possess some *Bos indicus* blood.

The Brahmans or Zebus that were introduced from India consisted of representatives of various breeds of *Bos indicus* which have existed for centuries. These breeds were usually named after the province in which they were developed.

The first *Bos indicus* cattle were imported into the United States in 1849 by James Bolton Davis of South Carolina, and later importations were made into Georgia, Louisiana, and Texas. The most important of these introductions were made by the Pierce Estate, Wharton County, Tex., in 1906, and by John T. Martin of San Antonio, Tex., (from Mexico) in 1924. Garcia Brothers of McAllen, Tex., and Hogue Poole of Cotulla, Tex., imported a number of Indu-Brazil cattle from Mexico in 1946. For a time Brahman cattle were confined to the Gulf Coast region, but since 1942 they have spread to 46 of the 48 States, and particularly to the North and West. The introduction of Brahman cattle into the United States was for the express purpose of crossing them with other breeds to develop new types with greater adaptation to the Gulf Coast regions. Probably no other type of cattle has shown such a phenomenal increase in recent years.

Brahman cattle have several characteristics which differentiate them from other types. These are a distinct hump over the shoulders, and

FIGURE 21.—Brahman bull.

an excess of loose skin under the throat, on the dewlap, and in the regions of the navel and sheath.

Desirable specimens have deep bodies with full muscle development in the hind quarters. The rump is usually slightly drooping, but nevertheless should be full and round. The ears are long and drooping. The color varies; however, the prevailing color is gray. The lighter grays are popular. Red of various shades is less frequent but equally acceptable. Other acceptable colors are gray or red with white spots, gray with odd brown or red spots, black with white spots, brown, or white. A brindle color is a disqualification for registration.

The better specimens of present day American Brahman cattle in the United States are markedly different from their Indian ancestors. The American Brahman represents the result of intermating the several Indian types brought to this country and making selections of the more beefy individuals that were adapted to our conditions.

Among American Brahmans the following families and sub-families are recognized: Imperator, Aristocrata, Moroto Invinca, Manso, Imperor, Resoto Manso, Aristocrat Manso, and Manimoso Manso.

The Indu Brazil Zebu is a composite breed created from the fusion of Indian Gir and Guzerat blood. The most prominent families of Indu Brazil cattle trace to the following sires: Repucho, Gaucho, Jacarando, Rio Negro, Brilhante, and Rio Pardo.

Brahmans, and particularly crossbred animals with Brahman blood, have given a good account of themselves in the feed lots of the South and Southwest.

CHAROLAISE

The Charolaise (other spellings Charollaise, Charollais) is one of the most important breeds of French cattle (fig. 22). It is white or light creamy colored and originated in the Province of Charollais. It is a popular breed in the prairie section of central France. It is one of the largest of all breeds of beef cattle.

FIGURE 22.—Charolaise bull.

FIGURE 23.—Beefmaster cow.

Only a very small number of Charolaise cattle have been brought into the United States. The breed is being used in Cuba in grading up commercial herds of native cattle. This breed appears to combine well with the Brahman and also the British breeds. It is being used by breeders particularly in Texas, Louisiana, and Florida who are practicing various systems of crossbreeding.

BREEDS DEVELOPED IN THE UNITED STATES
BEEFMASTER

In 1908, the late Edward C. Lasater of Falfurrias, Tex., who was the ninth president of the Texas and Southwestern Cattle Raisers' Association, began using Brahman sires on a large herd of Hereford and Shorthorn cattle. He used Gir, Nellore, and Guzerat Brahmans. His Herefords were selected for red pigmentation around the eyes and for milk production. (Fig. 23.)

After Mr. Lasater's death on March 20, 1930, his son, Tom Lasater, began to combine the entire breeding herd into what is now known as the "Lasater Beefmaster," a trade-mark which has been copyrighted through the United States Patent Office. It is estimated that the Beefmaster contains half Brahman, one-quarter Hereford, and one-quarter Shorthorn. Tom Lasater has stressed the six principal characteristics which his father selected for (1) disposition, (2) fertility, (3) weight, (4) conformation, (5) thriftiness, and (6) milk production. He has also dwelt on the progeny test.

While no specific color requirements have been established in the selection program, the majority of Beefmasters are dun, brown, reddish-brown, or red with some white extensions and spots in certain individuals.

As is to be expected from their genetic background, they are relatively large cattle.

FIGURE 24.—Brangus bull.

BRANGUS

The word "Brangus" is a registered trade name and can be used only as applied to cattle registered with the American Brangus Breeders Association. This Association was founded in Vinita, Okla., on July 29, 1949, by 54 breeders of this new type of cattle. Brangus cattle are three-eighths Brahman and five-eighths Angus. They are black and are hornless. (Fig. 24).

To produce a Brangus, an animal possessing one-quarter Brahman and three-quarters Angus may be bred to an animal that is one-half Brahman and one-half Angus, or an animal that is three-quarters Brahman and one-quarter Angus may be bred to an Angus.

Brangus are the only cattle that are registered in the permanent register of the Association.

Foundation stock for Brangus cattle may be enrolled with the Association. Such stock may include purebred Brahmans, purebred Aberdeen-Angus, and animals possessing either one-quarter, three-eighths, or one-half Brahman and the rest Aberdeen-Angus.

All stock to be registered as Brangus or enrolled as foundation stock have to be inspected by an Association appraisal committee.

CHARBRAY

The Charbray was developed from a cross of the Charolaise and the Brahman and shows some of the characteristics of each parental breed (Fig. 25). In order to qualify for registration in the herd book of the American Charbray Breeders' Association, animals must have at least one-eighth and not more than one-quarter Brahman and the remaining fraction must be Charolaise. The various Charbray blood lines have been developed for the most part from Charolaise bulls

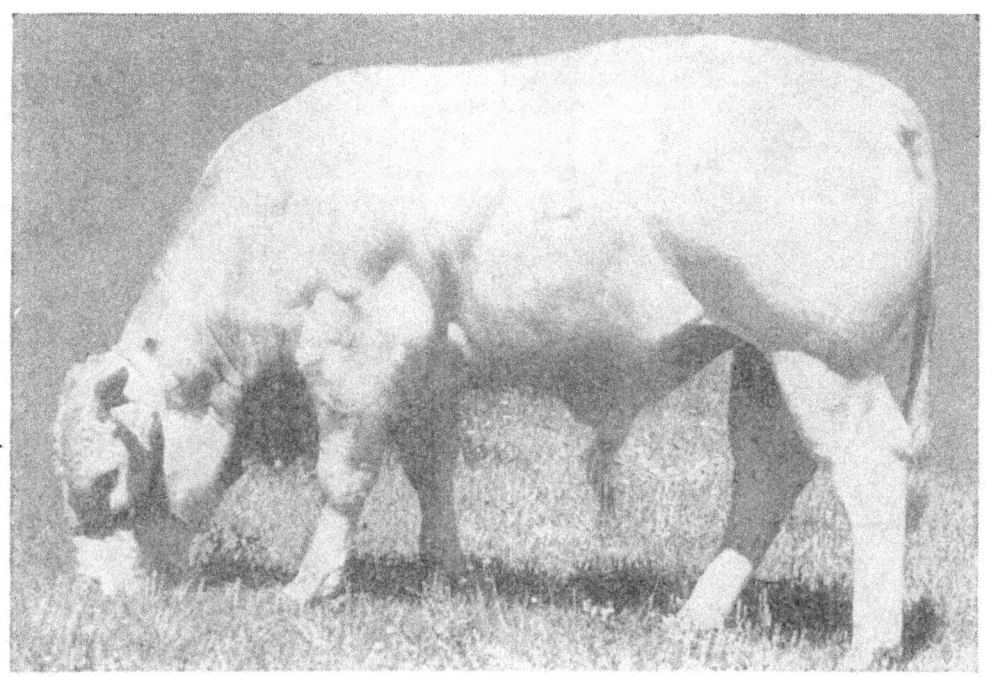

FIGURE 25.—Charbray yearling bull.

which were imported during the period 1936 to 1942 from the herd of the late Juan Pugibet of Mexico City.

In color markings, Charbray animals resemble the Charolaise in that they are usually a creamy white. They are horned and while they show almost no evidence of the Brahman hump they do show slight evidence of the Brahman dewlap and some excess skin in that region. They are large cattle and produce vigorous, fast-growing calves.

Because of their desirable grazing habits and ease of handling, they are being used by commercial operators in the South and West for crossing purposes.

SANTA GERTRUDIS

The Santa Gertrudis breed of beef cattle was developed on the King Ranch in Southeast Texas from a cross between beef type Shorthorn cows and beef type Brahman bulls. Crossbreeding between these breeds of different species, *Bos taurus* and *Bos indicus*, started in 1910 and continued for approximately 15 years. The breed first started with the bull Monkey. His sons and grandsons were used in the herds as rapidly as possible until eventually only bulls which descended from him were used. After nearly 30 years of following a well-defined plan of selection, inbreeding, and linebreeding of the offspring of Monkey, the desired characteristics had become sufficiently well fixed that the Santa Gertrudis was generally recognized as a new and distinct breed by 1940. It is the first distinctly American breed of cattle to be recognized. Its name is derived from the Santa Gertrudis division of the King Ranch, where the original foundation cattle were maintained and the breed was developed.

The Santa Gertrudis is a large beef animal with mature cows frequently attaining 1,600 pounds and mature bulls 2,000 pounds in weight

FIGURE 26.—Santa Gertrudis cow.

on pasture (fig. 26). It is solid cherry-red in color with flesh colored mucosae. It is a horned breed, although polled individuals are sometimes seen. The ears are somewhat large and semi-pendulant. Both males and females are loose-hided, showing considerable dewlap and underline skin folds. A short, straight hair is characteristic.

All Santa Gertrudis cattle are descendants of Monkey, the foundation sire of the breed. There are three families or sire lines of breeding, descendants of the bulls Tipo, Pop Eye, and Miguel. All three were grandsons of Monkey. These families have been developed on the King Ranch and are currently being maintained as distinct lines of breeding. Of these three families, the Tipo is the most prominent.

The King Ranch established a herd on its property near Lexington, Ky., for the purpose of developing, through selection, a line of breeding showing specific adaptation to the cooler climatic belt of the United States.

The Santa Gertrudis breed was developed purposely for adaptation to subtropical climates and semiarid ranching conditions. It makes large gains on grass and rustles for a living in areas of scant forage. Because of its heat tolerance, it thrives in climates too warm for optimum development of standard European breeds.

The Santa Gertrudis Breeders International was organized in 1951. The association is classifying all Santa Gertrudis cattle in preparation for the registration of acceptable individuals.

Santa Gertrudis cattle have been exported to Cuba, to most of the Central and South American countries, the Territory of Hawaii, and Australia.

U. S. GOVERNMENT PRINTING OFFICE: 1959

www.ingramcontent.com/pod-product-compliance
Lightning Source LLC
Chambersburg PA
CBHW060008230526
45472CB00008B/1999